GLOBES

by Robin Nelson

first step nonfiction

Lerner Publications · Minneapolis

What does a globe show?

TURKEY
IRAN
AFGHANISTAN
PAKISTAN
INDIA
CHINA
EGYPT
SAUDI
ARABIA
OMAN
YEMEN
SUDAN
CHAD
ETHIOPIA
SOMALIA
KENYA
ARABIAN SEA
BAY OF BENGAL
SRI LANKA
MALDIVES
SEYCHELLES
MAURITIUS
MADAGASCAR
Equator
INDIAN

Earth

NORTH AMERICA
UNITED STATES
HUDSON BAY
MANITOBA
ONTARIO
ALBERTA
TEXAS
New York
Philadelphia
Montreal
Dallas
Houston
Los Angeles
San Francisco
Chicago
Detroit
GULF OF MEXICO
CUBA
MEXICO
Mexico City
CENTRAL AMERICA
VENEZUELA
COLOMBIA
ECUADOR
SARGASSO SEA
Tropic of Cancer

Continents

NORTH ATLANTIC OCEAN
LABRADOR SEA
Cape Farewell
LABRADOR BASIN
NEWFOUNDLAND LABRADOR
QUEBEC
Battle Harbour
Str. of Belle Isle
ISLAND OF NEWFOUNDLAND
Gander
St. John's
Cape Race
St. Pierre & Miquelon Is. Fr.
NEWFOUNDLAND BASIN
Grand Banks of Newfoundland
NOVA SCOTIA
Halifax
Charlottetown
Sydney
Boston, MASS.
Concord, N.H.
Augusta
Providence R.I.
Hartford, CONN.
Gulf Stream
North Atlantic Current
UNITED KINGDOM
SCOTLAND
N. IRELAND
IRELAND
Dublin
Cork
WEST EUROPEAN BASIN
50°
AZORES
Corvo
Flores
Graciosa
Faial
Terceira
Pico 7713
São Miguel
Ponta Delgada
Santa Maria
Azores (Açores) Portugal
IBERIAN BASIN
A Coruña
Vigo
Oporto
PORTUGAL
NORTH AMERICAN BASIN
SARGASSO SEA
Great Meteor Tablemount
CANARY BASIN
Madeira
Funchal
30°
Canary Islands Spain
La Palma
Hierro
Tenerife
Las Palmas
NARES PLAIN
PUERTO RICO TRENCH
Tropic of Cancer
Canary Current
Ad Dakhla
Nouadhibou
Cap Blanc
Nouakchott
ST. KITTS AND NEVIS
ANTIGUA AND BARBUDA
DOMINICA
ST. LUCIA
Guadeloupe Fr.
Martinique Fr.
Lesser Antilles
CAPE VERDE
Santo Antão
São Vicente
São Nicolau
Sal
Boa Vista
Maio
São Tiago
Praia
Dakar
St.-Louis

Oceans

GABON
CONGO
Brazzavill
Kinshasa
DEM. REP. OF
THE CONGO
Inongo
Lake Mai-Ndombe
Kindu
Bukavu
Kigali
RWANDA
Lake Victoria
Nairobi
Mt. Kenya
Bujumbura
BURUNDI
Mwanza
Kilimanjaro
Malindi
Mombasa
Kigoma
Tabora
Tanga
TANZANIA
Pemba Island
Zanzibar
Dodoma
Dar es Salaam
Iringa
Mafia Island
Mbeya
Kalemie
Lake Tanganyika
Kasai
Pointe-Noire
Boma
Matadi
Angola
Cabinda
Soyo
Tshikapa
Kabalo
Kamina
Luanda
Malanje
KATANGA
Mbala
L. Mweru
Kasama
Lindi
Mtwara
Cabo Delgado
Aldabra Is.
ANGOLA
Dilolo
Kolwezi
Likasi
Lubumbashi
Lobito
Benguela
Kuito
Huambo
Luena
Kitwe
Ndola
6070
MALAWI
Lake Malawi
Lichinga
Moroni
COMOROS
Pemba
Nacala
Mozambique
Namibe
Lubango
Mongu
Kabwe
Lusaka
ZAMBIA
Lilongwe
Zambezi
Zomba
Tete
Nampula
MOZAMBIQUE
MADAGASCAR
Xangongo
Kunene
Cape Fria
Livingstone
L. Kariba
Victoria Falls
Mutarara
Harare
ZIMBABWE
Mozambique Channel
Current
Grootfontein
Maun
Beira
Bulawayo
Francistown
Brandberg
8442
Windhoek
BOTSWANA
Kalahari Desert
Walvis Bay
NAMIBIA
Thohoyandou
Pietersburg
6980
Limpopo
Ponta da Barra
Inhambane
Toliara
Antananarivo
Antsirabe
Fianarantsoa
6421
Gaborone
Pretoria
Lüderitz
Keetmanshoop
Johannesburg
Maputo
Mbabane
SWAZILAND
Agulhas
Cap Ste. Marie
Upington
Kimberley
SOUTH AFRICA
Port Nolloth
Orange
Maseru
Pietermaritzburg
Bloemfontein
Durban
LESOTHO
NATAL
MADAGASCAR PLATEAU
Calvinia
De Aar
Lambert's Bay
Benguela Current
Queenstown
Worcester
Bisho
East London
Cape Town
George
Port
Mossel Bay
Cape of Good Hope
Cape Agulhas
CAPE BASIN
BASIN

Countries

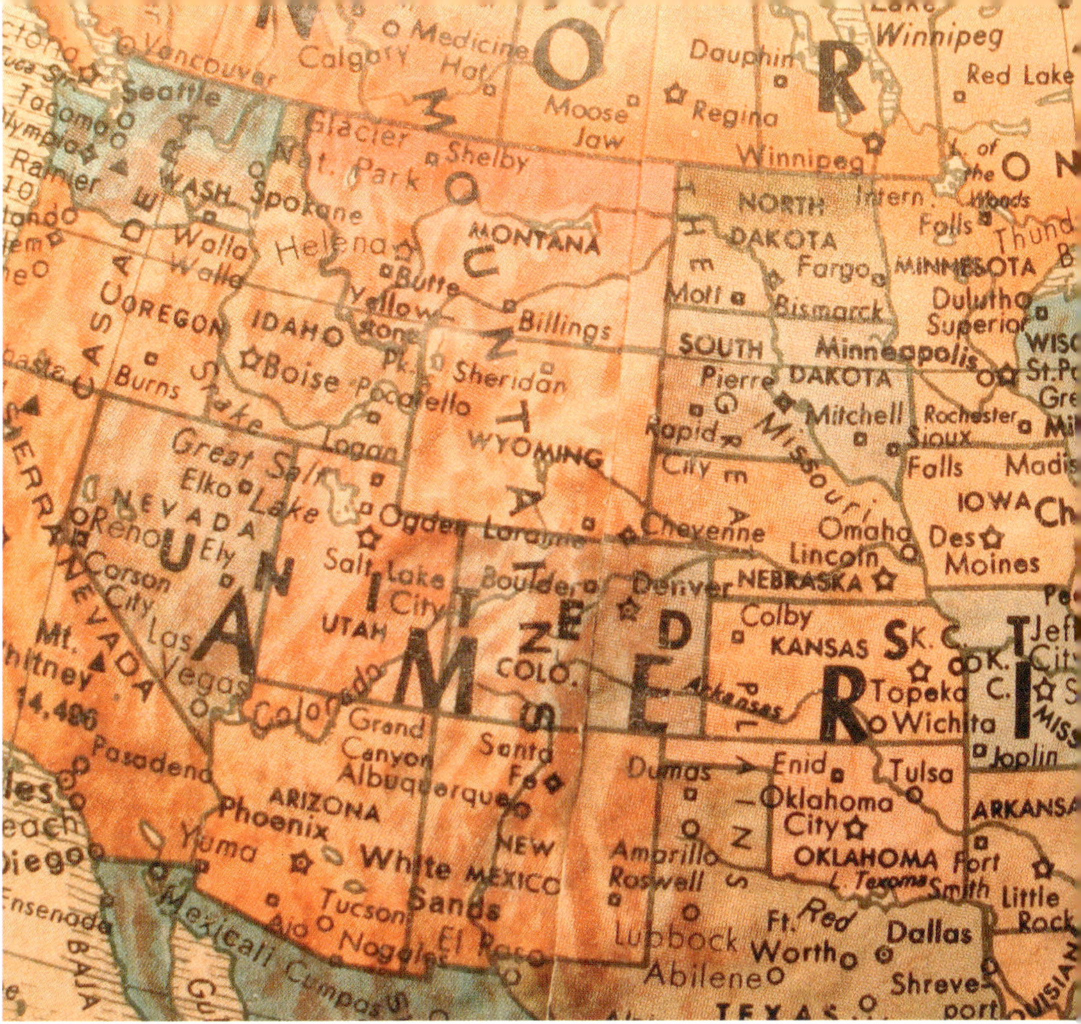
Medicine Hat
Calgary
Vancouver
Seattle
Tacoma
Olympia
Rainier
Dauphin
Winnipeg
Red Lake
Moose Jaw
Regina
Winnipeg
Glacier Nat. Park
Shelby
WASH.
Spokane
Walla Walla
Helena
MONTANA
Butte
NORTH DAKOTA
Fargo
MINNESOTA
Mott
Bismarck
Duluth
Superior
Intern. Falls
L. of the Woods
OREGON
IDAHO
Yellowstone
Billings
Boise
Burns
Pocatello
Sheridan
SOUTH DAKOTA
Pierre
Minneapolis
Mitchell
Rochester
Rapid City
Missouri
Sioux Falls
Snake
Logan
WYOMING
Great Salt Lake
Elko
NEVADA
Reno
Ogden
Laramie
Cheyenne
Omaha
Lincoln
IOWA
Des Moines
Ely
Carson City
Salt Lake City
Boulder
Denver
NEBRASKA
Colby
KANSAS
K.C.
UTAH
COLO.
Las Vegas
Mt. Whitney
Topeka
Wichita
Arkansas
Colorado
Grand Canyon
Santa Fe
Albuquerque
Pasadena
Dumas
Enid
Tulsa
Joplin
Oklahoma City
ARIZONA
Phoenix
Yuma
White Sands
NEW MEXICO
Amarillo
OKLAHOMA
Fort Smith
Roswell
L. Texoma
Little Rock
Tucson
Ajo
Nogales
El Paso
Lubbock
Ft. Worth
Red
Dallas
Abilene
Shreveport
Ensenada
BAJA
Mexicali
TEXAS

Mountains

Itacoatiara
Negro
Óbidos
Almeirim
Manaus
Amazon
Coari
Borba
Tapajós
Belterra
São
Francisco
Manicoré
Curupachi
Xingu

Rivers